Leckie
the education publisher
for Scotland

National 5
Chemistry
Lab Skills

for SQA assessment

Bob Wilson

© 2019 Leckie

001/24012019

10 9 8 7 6 5 4 3

ISBN 9780008329648

Published by
Leckie
An imprint of HarperCollinsPublishers
Westerhill Road, Bishopbriggs, Glasgow, G64 2QT
T: 0844 576 8126 F: 0844 576 8131
leckiescotland@harpercollins.co.uk   www.leckiescotland.co.uk

HarperCollinsPublishers
Macken House, 39/40 Mayor Street Upper,
Dublin 1, D01 C9W8, Ireland

A CIP Catalogue record for this book is available from the British Library.

Publisher: Sarah Mitchell
Commissioning editor: Gillian Bowman
Managing Editor: Craig Balfour

Special thanks to
Jouve India (layout, illustration and project managment)
Dr Jan Schubert (proofreading)
Jess White (copyediting)

Printed and bound in the UK using 100% Renewable Electricity at CPI Group (UK) Ltd

This book contains FSC™ certified paper and other controlled sources to ensure responsible forest management.

For more information visit: www.harpercollins.co.uk/green

Contents

How to use this book

Scientific skills are important for any science qualification. Your SQA National 5 Chemistry course develops the following skills and allows you to become familiar with the the techniques and analytical methods listed. These can be tested in your **examination**.

Skills
Planning; selecting; presenting; processing; concluding; predicting; evaluating.

Techniques and analytical methods
Monitoring the rate of reaction; making a standard solution and using it in a titration; identifying ions; making electrochemical cells; making a soluble salt; determination of E_h; electrolysis of an ionic solution.

Underlying chemistry
These introductory statements cover the underlying chemistry which is needed to understand the context of the experiment and could help with any Assignment based on it.

Learning outcomes
This is a summary of the skills you will develop during each experiment.

Aim
A clear statement of the aim of the experiment is given. Any conclusion must be based on this aim.

Apparatus list
Your teacher will ensure that all the apparatus you need for the experiment can be found in the laboratory. You can use this list to check that you have everything you need to start your work. Items listed in **bold** are those you need to be familiar with the use(s) of for your examination.

Safety notes
You should be aware of safety when carrying out an experiment. These notes will help you be aware of any safety issues! Your teacher will advise on safety information for each experiment, so pay attention.

Experimental precautions
We've included some precautions which will help to make your experiment more accurate, valid and reliable.

Method
Always make sure you read every step of the method before you begin work. This will help you avoid mistakes and will give you an idea of what outcomes to look for as you complete each step.

Record your results
For each experiment there is a place to record the results of your work. Make sure you keep your data tables, graphs, answers and notes clear and neat.

Check your understanding
This section ensures you understand the outcomes from each experiment.

Exam-style question
These questions give you practice with the type of question you may get in the final exam.

Assignment support
For your assignment, you must carry out an experiment and collect data for use in your report. This section gives you some ideas of experiments you could carry out for your assignment.

Notes
These pages can be used for making extra notes when doing experiments or for doing calculations.

1 Monitoring the rate of reaction

Underlying chemistry

The rate of a chemical reaction can be monitored by measuring the volume of gas produced over time. If the gas produced has low solubility it can be collected over water. Changing variables such as concentration, temperature or particle size will affect the rate of a reaction and this can be measured.

Learning outcomes

- Monitor the rate of a chemical reaction and describe the changes observed.
- Monitor the rate of a chemical reaction when the concentration of one reactant is changed and describe any changes observed.
- Draw line graphs of volume of gas against time for different concentrations of a reactant and explain any differences.
- Collect a gas over water.

Aim

To monitor the effect of changing the concentration of the acid on the rate of the reaction between marble chips (calcium carbonate) and hydrochloric acid.

Apparatus list

Your teacher will ensure that all the apparatus you need for the experiments can be found in the laboratory. You can use this list to check that you have everything you need.

- 100 cm^3 **measuring cylinder**
- 10 cm^3 measuring cylinder
- **boiling tube**
- boiling tube rack
- rubber bung with **delivery tube**

- plastic tub
- 1·0 mol l^{-1} hydrochloric acid
- 2·0 mol l^{-1} hydrochloric acid
- bottle of large marble chips
- timer

Safety notes

Your teacher will go over general safety rules with you and precautions which should be taken when carrying out these experiments.

In particular you should be aware of the following:

- Hydrochloric acid is an irritant.

Experimental precautions

- Make sure the measuring cylinder is full of water before you invert it in the water in the plastic tub.
- Take care not to clamp the measuring cylinder too tightly.
- As soon as you add the acid to the boiling tube, place the stopper in the boiling tube and start the timer.

Method

1. Place the boiling tube in the boiling tube rack.
2. Select two large marble chips of similar size.
3. Add the marble chips to the boiling tube.
4. Fill the 100 cm³ measuring cylinder with water.
5. Set up the boiling tube, delivery tube, measuring cylinder full of water and plastic tub of water as shown in the diagram.

Figure 1.1

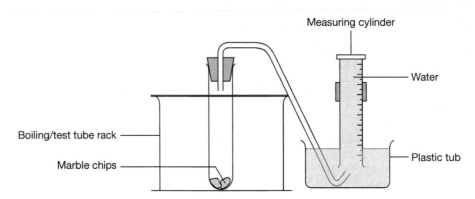

(**Note:** Your teacher may demonstrate how to invert the measuring cylinder in the tub of water.)

6. Measure out 10 cm³ of 1·0 mol l⁻¹ hydrochloric acid in the small measuring cylinder.
7. Add the acid to the boiling tube, reconnect the delivery tube and start the timer.
8. Note the volume of gas produced every minute for 10 minutes. (You can stop taking readings if the volume of gas remains unchanged before 10 minutes.)
9. Repeat the experiment using 10 cm³ of 2·0 mol l⁻¹ hydrochloric acid.

Record your results

1. Present your results in a table with appropriate headings and units of measurement. You will need separate tables for each concentration of hydrochloric acid.

2. Collect a piece of graph paper. Draw line graphs of your results – both sets of results should be on the same graph so a comparison can be made. Make sure you use a suitable scale, in order to fill most of the graph paper, and your axes are clearly labelled and units are included.

   Stick your graph paper here.

3. Use the relationship rate $= \frac{\Delta \text{quantity}}{\Delta t}$ (which can be found in the SQA data booklet) and the information in the graphs to:

   a. Calculate the average rate of reaction (in $cm^3\,min^{-1}$) between 1 and 2 minutes when $1{\cdot}0$ mol l^{-1} hydrochloric acid is used.

   b. Calculate the average rate of reaction (in $cm^3\,min^{-1}$) between 1 and 2 minutes when $2{\cdot}0$ mol l^{-1} hydrochloric acid is used.

4. State what conclusion can be reached regarding the effect changing the concentration has on the rate of reaction.

   ....

   ....

Check your understanding

1. Explain the importance of using the same mass of marble chips and volume of acid in each reaction.

   ....

   ....

2. a. State what happens to the average rate of each reaction as the reactions progress.

      ....

   b. Explain your answer to part (a).

      ....

3. Explain why each graph eventually levelled off.

   ....

4. Write a balanced equation for the reaction.

   ....

5. Suggest one change which could be made to the experiment in order to improve the accuracy of your results.

   ....

Exam-style question

1. A student monitored the rate of reaction between excess magnesium ribbon and dilute hydrochloric acid (HCl), using a gas syringe to collect the gas produced.

Figure 1.2

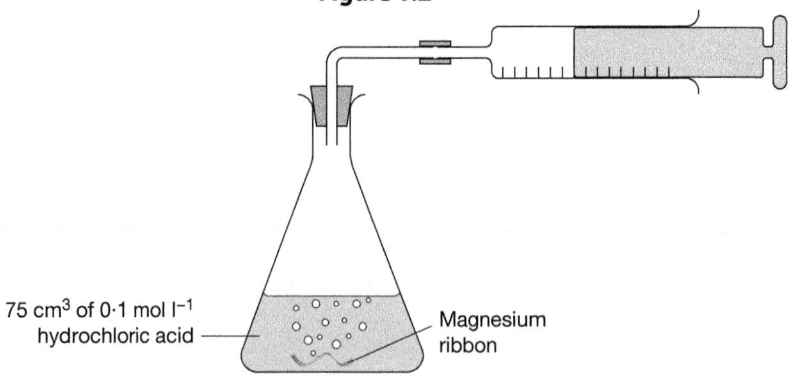

75 cm³ of 0·1 mol l⁻¹ hydrochloric acid

Magnesium ribbon

a. State what is meant by 'excess magnesium'. [1 mark]

....

b. Describe a chemical test the student could carry out to prove that hydrogen gas was produced. [2 marks]

....

....

c. The student obtained the results shown.

Time (s)	0	10	20	30	40	50	60
Volume of gas (cm³)	0	46	59	70	75	78	78

   i. Calculate the average rate of reaction between 10 and 30 seconds.
   Your answer must include the appropriate units.

   Show your working clearly. [3 marks]

   ii. The student carried out a similar experiment using 75 cm³ of 0·1 mol l⁻¹ sulfuric acid, $H_2SO_4(aq)$. The volume of gas collected after 50 s was 156 cm³.

   Explain why there was a greater volume of gas produced. [1 mark]

   ....

d. State how the rate of the reaction would be affected if the experiment was repeated using powdered magnesium. [1 mark]

....

Assignment support

The effect on the rate of reaction of changing variables such as temperature and particle size could be investigated. The effect of adding a catalyst on the rate of reaction could be investigated. For example, the effect of adding copper or copper ions to the reaction between zinc and hydrochloric acid.

2(a) Making a standard solution

Underlying chemistry

Solutions of accurately known concentration are known as standard solutions. A standard solution of a base can be used in a titration to determine, accurately, the volume needed to neutralise an acid.

Learning outcomes

- Use a balance to weigh out accurately a mass of solute.
- Make a standard solution.
- Calculate the concentration of the standard solution.

Aim

To make a standard solution of sodium hydroxide to be used in a titration to determine the concentration of an acid.

Apparatus list

Your teacher will ensure that all the apparatus you need for the experiment can be found in the laboratory. You can use this list to check that you have everything you need.

- 100 cm³ volumetric flask and stopper
- **electronic balance**
- 100 cm³ **beaker**
- **filter funnel**
- **filter paper**

- wash bottle with deionised water
- glass stirring rod
- solid sodium hydroxide
- spatula

Safety notes

Your teacher will go over general safety rules with you and precautions which should be taken when carrying out this experiment.

In particular you should be aware of the following:

- Sodium hydroxide is very corrosive and should not be handled.
- Heat is produced when solid sodium hydroxide is dissolved in water.
- Never add water to the solid sodium hydroxide.

Experimental precautions

- The more accurate you are in measuring the mass of the sodium hydroxide the more accurate the concentration of your standard solution will be.
- When filling the volumetric flask, always ensure the bottom of the meniscus is level with the line on the neck of the flask.

Method

1. Add approximately 50 cm³ of deionised water to the beaker – use the markings on the side of the beaker to judge the volume.

2. Switch on the balance and place the filter paper on it.

3. Press the tare button to reset the reading to zero.

Figure 2.1

4. Weigh out approximately 0·4 g of sodium hydroxide onto the filter paper on the balance – use a spatula.

   (The important thing here is to weigh out as close to 0·4 g as you can and record accurately the weight shown on the balance – it does not have to be exactly 0·4 g.)

Figure 2.2

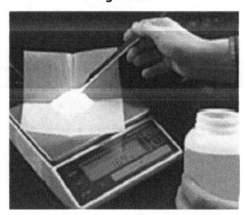

5. Carefully pour the sodium hydroxide from the filter paper into the 50 cm³ of deionised water in the beaker.

6. Stir the mixture with the glass rod until all of the sodium hydroxide has dissolved.

7. Place the filter funnel into the volumetric flask and transfer the sodium hydroxide solution into the flask.

8. Use the wash bottle to rinse all of the solution from the beaker.

9. Use the wash bottle to rinse the filter funnel.

10. Use the wash bottle to add deionised water to the line on the volumetric flask.

11. Put the stopper on the flask and gently invert it (turn it upside down) several times.

Important: Keep your standard solution for doing the titration in experiment 2(b).

Record your results

Calculate the concentration of the sodium hydroxide solution by first calculating the number of moles of sodium hydroxide using the mass you weighed out. Then use the number of moles to calculate the concentration of the sodium hydroxide solution.

Use the following relationships, which can be found in the SQA data booklet:

$$\text{Moles} = \frac{\text{mass}}{\text{gram formula mass}}$$

$$\text{Concentration} = \frac{\text{moles}}{\text{volume (l)}}$$

Check your understanding

1. Explain why the mass of sodium hydroxide weighed out need be only approximately 0·4 g but the actual mass has to be recorded accurately.

   ....

   ....

2. Explain why the tare facility on the balance is useful when weighing out the sodium hydroxide.

   ....

   ....

3. Suggest why the beaker and the filter funnel have to be rinsed in this way.

   ....

   ....

4. Suggest why tap water is not used to rinse the beaker and filter funnel.

   ....

   ....

Exam-style question

1. A student wanted to make up a standard solution of sodium carbonate (Na_2CO_3: GFM = 106 g). They weighed out 5·3 g of solid sodium carbonate and made up a 500 cm³ solution in a volumetric flask.

   a. State what is meant by a standard solution. [1 mark]

   ....

   b. Calculate the concentration of the standard solution the student made.
   Your answer must include the appropriate units.
   Show your working clearly. [3 marks]

2(b) Using a standard solution in a titration

Underlying chemistry

A standard solution of a base can be used in a titration to determine, accurately, the volume needed to neutralise an acid. The volume of the acid can then be used to calculate its concentration. An indicator is used to show when the endpoint is reached. Titre volumes within 0.2 cm^3 are considered concordant.

Learning outcomes

- Carry out a titration using a standard solution and an acid of unknown concentration.
- Calculate the concentration of the acid.

Aim

To determine the concentration of a solution of hydrochloric acid by titration with a standard sodium hydroxide solution.

Apparatus list

Your teacher will ensure that all the apparatus you need for the experiment can be found in the laboratory. You can use this list to check that you have everything you need.

- the standard solution of sodium hydroxide you made in part (a)
- hydrochloric acid of unknown concentration
- phenolphthalein indicator
- 50 cm^3 **burette**
- 20 cm^3 **pipette and safety filler**
- 250 cm^3 **conical flask**
- white tile

Safety notes

Your teacher will go over general safety rules with you and precautions which should be taken when carrying out this experiment.

In particular you should be aware of the following:

- Sodium hydroxide is very corrosive.
- Always use a safety filler with the pipette.
- Hydrochloric acid is an irritant.

Experimental precautions

- The more accurate you are when measuring the volume of sodium hydroxide using the pipette the more accurate your final result will be.
- When filling the pipette always ensure the bottom of the meniscus is level with the line on the pipette.
- The more accurate your reading of the burette is the more accurate your final result will be.
- When taking readings from the burette you should always take the reading from the bottom of the meniscus.

Method

1. Use a pipette (with safety filler) to transfer 20 cm³ of the standard sodium hydroxide solution into the conical flask.

   (**Note**: Your teacher may give you a different sodium hydroxide solution to the one you prepared.)

2. Add a few drops of phenolphthalein indicator to the solution and note the colour.

3. Add the hydrochloric acid to the burette until it gets near the zero mark. Note the volume on the burette.

4. Place the conical flask on a white tile under the burette.

   Your apparatus should be set up as shown in the diagram. A clamp should be used to hold the burette.

Figure 2.3

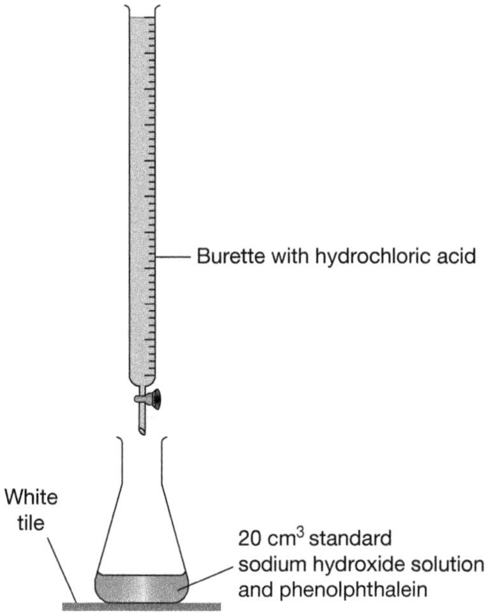

Burette with hydrochloric acid

White tile

20 cm³ standard sodium hydroxide solution and phenolphthalein

5. Add the acid to the alkali while swirling the conical flask.

6. Stop adding acid immediately you see the indicator colour disappear.

   Note the colour of the indicator.

   Note the volume on the burette.

7. Work out the volume of acid added and note it.

8. Repeat the titration using a fresh sample of the sodium hydroxide solution. This time slow down the rate at which you add the acid when you get within 2 cm³ of your first titration. Add the acid drop by drop when you are near the expected endpoint.

9. Repeat the titration until you get two volumes which are within 0·2 cm³ of each other.

Record your results

1. State the colour change observed when the endpoint is reached.

....

2. Present your results in the form of a table with appropriate headings and units.

3. Calculate the average (mean) titre.

4. The balanced equation for the reaction is:

NaOH(aq) + HCl(aq) → NaCl(aq) + $H_2O(\ell)$

Calculate the concentration of the hydrochloric acid (C_1) using the relationship:

$$\frac{C_1 V_1}{n_1} = \frac{C_2 V_2}{n_2}$$ (This relationship can be found in the SQA data booklet.)

where C_1 = concentration of HCl(aq) (unknown)

V_1 = average titre (calculated in part 3)

n_1 = number of moles of acid from balanced equation

C_2 = concentration of standard sodium hydroxide solution (from part (a))

V_2 = volume of standard sodium hydroxide solution used

n_2 = number of moles of alkali from balanced equation

Show your working clearly and include units.

Check your understanding

1. Explain why it is not necessary to have the volume of acid in the burette at zero at the start of each titration.

....

2. Explain why it is important to swirl the conical flask during the titration.

....

3. Suggest why a white tile is placed under the conical flask during the titration.

....

4. State what has happened when the endpoint is reached.

....

5. Suggest why universal indicator would not be suitable for the titration.

....

....

6. State why the first titre should not be used when calculating the average titre.

....

....

Exam-style questions

1. A student carried out titrations to find out the volume of nitric acid needed to neutralise 20 cm^3 of 0·05 mol l^{-1} potassium hydroxide solution. The volume can then be used to calculate the concentration of the acid. The results are shown in the table.

Titration	Volume of acid added (cm³)
1	13·9
2	13·6
3	13·3
4	13·1

a. The average volume of nitric acid that should be used to calculate the acid concentration is 13·2 cm^3. Explain why only the results of titration 3 and titration 4 are used to calculate this average. [1 mark]

....

b. Calculate the concentration of the nitric acid.

$KOH(aq) + HNO_3(aq) \rightarrow KNO_3(aq) + H_2O(\ell)$

Your answer must include the appropriate units.

Show your working clearly. [3 marks]

2. The diagram shows the assembled apparatus used to find the volume of hydrochloric acid needed to neutralise a solution of sodium carbonate.

Figure 2.4

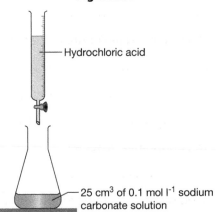

Hydrochloric acid

25 cm³ of 0.1 mol l⁻¹ sodium carbonate solution

$$Na_2CO_3(aq) + 2HCl(aq) \rightarrow 2NaCl(aq) + H_2O(\ell) + CO_2(g)$$

a. State what needs to be added to the flask so that the endpoint can be detected. [1 mark]

....

b. Name the piece of apparatus used to deliver the acid into the flask containing the sodium carbonate. [1 mark]

....

c. The average volume of acid needed to neutralise the sodium carbonate is 24·8 cm³.

Calculate the concentration of the hydrochloric acid.

Your answer must include the appropriate units.

Show your working clearly. [3 marks]

Assignment support

Titration can be used to analyse samples of wine or fruit juice to find how acidic they are.

All wines contain a certain amount of acid, mainly tartaric acid, which contributes to the taste of the wine. Winemakers have to check the total acidity of the wine batch to ensure it is within certain limits. Fruit juices contain citric acid. The concentration of the acid in a sample of wine or fruit juice can be calculated by titrating a sample with standard sodium hydroxide solution. A pH meter could be used to detect the endpoint instead of a chemical indicator.

3(a) Identifying ions – flame tests

Underlying chemistry

The identification of metals and non-metals in a compound is important in chemical analysis. Many metal ions produce colours when samples are held in a Bunsen flame and the specific colours can be used to help identify the metal. This is particularly the case for group 1 metals.

Learning outcomes

- Carry out flame tests on known metal compounds and observe the colours obtained.

Aim

To identify the metal ion present in a compound from the colour of flame produced when a sample is held in a Bunsen flame.

Apparatus list

Your teacher will ensure that all the apparatus you need for the experiments can be found in the laboratory. You can use this list to check that you have everything you need.

- Bunsen burner
- **test tube**
- test tube rack
- dimple tray
- spatula

- nichrome wire mounted on a handle or opened out paper clips
- powdered chlorides: calcium, copper, potassium and sodium
- 1.0 mol l^{-1} hydrochloric acid – for cleaning the nichrome wire
- Wash bottle with deionised water

Safety notes

Your teacher will go over general safety rules with you and precautions which should be taken when carrying out these experiments.

In particular you should be aware of the following:

- 1.0 mol l^{-1} hydrochloric is an irritant.
- The powdered chlorides are irritants.

Experimental precautions

- The Bunsen flame must be blue.
- The nichrome wire must be cleaned thoroughly before each test.
  (If you are using paper clips then a fresh one can be used for each test.)
- Sometimes the flame colour is difficult to see because there can be a yellow flame in addition to the colour of the metal ion. If this is the case the flame can be viewed through cobalt blue glass, which absorbs the yellow colour.

Method

1. Use a spatula to place a small amount of the chloride to be tested into a dimple on a dimple tray.
2. Half fill a test tube with hydrochloric acid and put it in the test tube rack.
3. Light the Bunsen burner and open the air hole until a blue flame is obtained.
4. Hold the nichrome wire in the flame for a few seconds.
5. Remove the wire from the flame and dip it into the test tube containing the hydrochloric acid.
6. Rinse the wire in deionised water.
7. Dip the wire in the metal chloride to be tested.
8. Hold the nichrome wire in the Bunsen flame, at the side of the blue cone.
9. Note the colour observed and record it, beside the name of the compound, in your results table.
10. You can repeat the test if you are not sure of the colour of the flame produced.
11. Repeat steps 4–9 for the other metal chlorides to be tested.

Figure 3.1

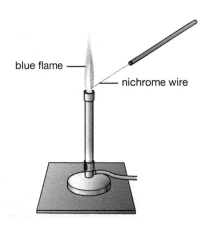

blue flame —

— nichrome wire

Record your results

Present your results in a table with appropriate headings.

Check your understanding

1. Suggest why a blue flame needs to be used.

   ....

2. Suggest why the nichrome wire should be cleaned thoroughly between tests.

   ....

3. State why the nichrome wire should be dipped in deionised water and not tap water.

   ....

   ....

Exam-style question

You may wish to use the SQA data booklet to help you answer this question.

1. a. Name the metal that forms compounds which give a green flame colour. [1 mark]

    ....

    b. Both lithium and strontium compounds produce similar red flames.
       Suggest how you could determine whether an unknown compound which produced
       a red flame contained lithium or strontium. [2 marks]

    ....

    ....

3(b) Identifying ions – forming a precipitate

Underlying chemistry

Some metal and non-metal ions can produce coloured precipitates when reacted with certain reagents and this can be used to help identify the ion in a compound.

Learning outcomes

- React known compounds with a reagent and observe the colour of the precipitate formed.

Aim

To identify the halide (group 7) ion present in a compound from the colour of the precipitate formed when it reacts with silver nitrate solution.

Apparatus list

Your teacher will ensure that all the apparatus you need for the experiments can be found in the laboratory. You can use this list to check that you have everything you need.

- three **test tubes**
- test tube rack
- **dropper**

- solutions of the following: sodium chloride, sodium bromide and sodium iodide
- 0.05 mol l^{-1} silver nitrate solution
- 0.1 mol l^{-1} nitric acid

Safety notes

Your teacher will go over general safety rules with you and precautions which should be taken when carrying out these experiments.

In particular you should be aware of the following:

- Silver nitrate is an irritant and will stain your skin or clothes if it comes in contact with them.

Experimental precautions

- It can sometimes be difficult to tell apart the colours produced with bromides and iodides. It is easier to tell the difference if they are side by side. If you still cannot see any difference tell your teacher and they may allow you to carry out an additional test to tell them apart.
- Make sure you label the test tubes so you remember which solution you are testing.

Method

1. Pour about 1 cm depth of each of the solutions to be tested into separate test tubes and place them side by side in the test tube rack.

2. Add a few drops of nitric acid to each test tube.

3. Use a dropper to add drops of silver nitrate solution to each test tube. Stop adding the silver nitrate when you can clearly see the colour of the precipitate.

4. Note the colour seen in each test tube.

Record your results

Present your results in a table with appropriate headings.

Check your understanding

1. State what is meant by a 'precipitate'.

   ....

2. The nitric acid is added so that it reacts with other ions which may be in the solution. Suggest why this is done.

   ....

3. When the test is carried out with sodium fluoride no precipitate is formed. Suggest why this is.

   ....

4. **a.** The word equation for sodium chloride reacting with silver(I) nitrate is shown.

   sodium chloride + silver(I) nitrate → silver chloride + sodium nitrate

   Write the chemical equation for the reaction, including state symbols.

   ....

   b. Name the precipitate formed when silver nitrate reacts with sodium bromide.

   ....

3(c) Identifying the ions in an unknown metal halide

Learning outcomes

- Identify the metal ion in unknown compounds by carrying out a flame test.
- React an unknown compound with a reagent in order to form a precipitate which can be used to identify the non-metal ion in the compound.

Aim

Your teacher will give you an 'unknown' metal halide solution to test.

Carry out a flame test and a silver nitrate test on the sample in order to identify the ions in the compound.

Results and conclusion

Exam-style questions

Page 8 of the SQA National 5 data booklet lists some compounds and their solubility in water.

1. a. i. Lithium iodide forms a precipitate with silver(I) nitrate.

Name the precipitate. [1 mark]

....

ii. Select a lithium compound from the solubility table which is not a halide
but would form a precipitate when mixed with silver(I) nitrate. [1 mark]

....

b. i. Lithium iodide and lithium chloride solutions look the same and both form a precipitate with silver(I) nitrate.

Suggest then how the formation of a precipitate can still be used as a test to distinguish between the two compounds. [1 mark]

....

ii. Describe a test which could be used to show that the compounds both contain lithium and not some other group 1 metal. [2 marks]

....

....

2. When a few drops of sodium hydroxide solution are added to separate solutions containing copper ions, magnesium ions and aluminium ions a white precipitate is produced with each. If more sodium hydroxide is added there is no change with the solutions of calcium and magnesium ions but the precipitate formed with the aluminium ions disappears.

a. A student stated that this test could be used to identify calcium, magnesium and aluminium ions present in compounds.

Comment on their statement. [2 marks]

....

....

b. i. Balance the following equation:

$CuSO_4(aq) + NaOH(aq) \rightarrow Cu(OH)_2(s) + Na_2SO_4(aq)$ [1 mark]

....

ii. Name the precipitate. [1 mark]

....

c. Identify the spectator ions in the following ionic equation.

$Fe^{3+}(aq) + 3Cl^-(aq) + 3Na^+(aq) + 3OH^-(aq) \rightarrow Fe(OH)_3(s) + 3Na^+(aq) + 3Cl^-(aq)$

[1 mark]

....

d. Ions of group 1 metals form soluble hydroxides so reaction with sodium hydroxide cannot be used to identify them.

Suggest another test which could be used to identify them. [2 marks]

....

....

Assignment support

Chloride ions are present in drinking water and in certain concentrations can cause the water to have an unpleasant taste. Chloride ions form a white precipitate with silver nitrate solution. Samples of drinking water could be titrated with silver nitrate solution and an indicator to detect the endpoint. The concentration of chloride ions in different water samples could then be calculated.

4(a) Electrochemical cells – using metal electrodes

Underlying chemistry

An electrochemical cell converts chemical energy into electrical energy. In a simple electrochemical cell a metal dipped in a solution of its own ions (known as a half-cell) is connected to another metal dipped in a solution of its own ions. The metals are connected by a wire and the solutions are connected by an ion (salt) bridge to complete the circuit. Electrons produced as a result of a redox reaction flow through the connecting wire from one metal to another.

Learning outcomes

- Set up an electrochemical cell using different metals in solutions of their own ions.
- Write oxidation and reduction ion-electron equations for the reactions taking place in half-cells.
- Write a redox equation for the overall reaction taking place in an electrochemical cell.
- Work out the path and direction of electron flow in a chemical cell.

Aim

To produce electricity from a chemical reaction by connecting half-cells consisting of metal electrodes in solutions of their own ions.

Apparatus list

Your teacher will ensure that all the apparatus you need for the experiment can be found in the laboratory. You can use this list to check that you have everything you need.

- one **zinc electrode**
- one **copper electrode**
- 0·1 mol l^{-1} zinc sulfate solution
- 0·1 mol l^{-1} copper(II) sulfate solution
- two 100 cm^3 **beakers**

- two leads with crocodile clips
- one voltmeter (high resistance)
- filter paper strip
- 0·1 mol l^{-1} potassium nitrate solution

Safety notes

Your teacher will go over general safety rules with you and precautions which should be taken when carrying out this experiment.

In particular you should be aware of the following:

- Copper(II) sulfate and zinc sulfate are irritants.
- Wear gloves when handling the filter paper soaked in potassium nitrate solution.

Experimental precautions

- The metals can be attached to the sides of the beakers using the crocodile clips. Make sure the crocodile clips holding the metals are not in the solutions.
- Make sure the filter paper soaked in potassium nitrate is in contact with both solutions in the beakers.

Method

1. Add the copper sulfate to one beaker, up to the 75 cm³ (ml) mark.
2. Add the zinc sulfate to the other beaker, up to the 75 cm³ (ml) mark.
3. Attach each metal to a separate lead, using the crocodile clips.
4. Place the copper metal in the copper sulfate solution and clip it to the side of the beaker.
5. Place the zinc metal in the zinc sulfate solution and clip it to the side of the beaker.
6. Connect the leads attached to the metals to the voltmeter.
7. Note the reading on the voltmeter.
8. Soak the filter paper in the potassium nitrate solution.
9. Add the filter paper soaked in potassium nitrate across the two beakers, making sure it is in both solutions.
10. Note the reading on the voltmeter.

Your apparatus should be set up as shown in the diagram.

Figure 4.1

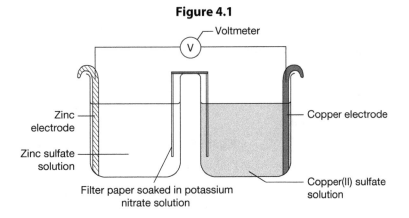

Record your results

Check your understanding

1. Explain the reason for the zero reading obtained before the filter paper was added across the two beakers.

    ....

2. Give the general term used to describe solutions like potassium nitrate which are electrically conducting solutions.

    ....

3. **a.** State the name given to the link which joins the solutions in the two beakers.

....

b. State what this link allows to happen.

....

4. Use the SQA data booklet to help you with these questions.

In the cell zinc metal atoms are oxidised and copper(II) ions are reduced.

a. Write an ion-electron equation for the oxidation of zinc.

....

b. Write an ion-electron equation for the reduction of copper(II) ions.

....

c. Write the redox equation for the reaction.

....

d. On the diagram, draw an arrow to show the path and direction of electron flow.

Exam-style question

1. A student set up the following cell.

Figure 4.2

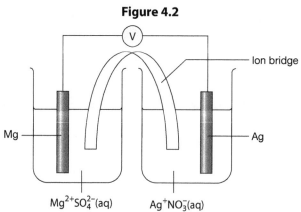

Ion bridge

Mg

Ag

$Mg^{2+}SO_4^{2-}(aq)$ $Ag^+NO_3^-(aq)$

a. On the diagram, draw an arrow to show the path and direction of electron flow.

You may wish to use the SQA data booklet to help you.   [1 mark]

b. Explain why an ion bridge is used to link the two beakers.   [1 mark]

....

c. Write the ion-electron equation for the reduction of the silver(I) ions.

You may wish to use the SQA data booklet to help you.   [1 mark]

....

d. State what would happen to the reading on the meter if the magnesium half-cell was replaced with a zinc half-cell.   [1 mark]

....

4(b) Electrochemical cells – using non-metal electrodes

Underlying chemistry

An electrochemical cell converts chemical energy into electrical energy. A cell can be made using non-metal electrodes, usually graphite (carbon) dipped into ionic solutions. The carbon electrodes are connected by a wire and the solutions are connected by an ion (salt) bridge to complete the circuit. Electrons produced as a result of a redox reaction flow through the connecting wire from one carbon rod to another.

Learning outcomes

- Set up an electrochemical cell using non-metal electrodes.
- Write oxidation and reduction ion-electron equations for the reactions taking place in half-cells.
- Write a redox equation for the overall reaction taking place in an electrochemical cell.
- Work out the path and direction of electron flow in a chemical cell.

Aim

To produce electricity from a chemical reaction by connecting half-cells consisting of non-metal electrodes.

Apparatus list

Your teacher will ensure that all the apparatus you need for the experiment can be found in the laboratory. You can use this list to check that you have everything you need.

- two **carbon electrodes**
- 0.1 mol l^{-1} iodine solution
- 0.1 mol l^{-1} sodium sulfite solution
- two 100 cm³ **beakers**
- two leads with crocodile clips
- one voltmeter (high resistance)
- filter paper strip
- 0.1 mol l^{-1} potassium nitrate solution

Safety notes

Your teacher will go over general safety rules with you and precautions which should be taken when carrying out this experiment.

In particular you should be aware of the following:

- The iodine solution will stain your skin and clothes if it comes in contact with them.

Experimental precautions and method

The experimental precautions and method are the same as for making cells using metals in part (a). The carbon (non-metal) rods are used instead of metals – one rod in each solution.

Record your results

Check your understanding

Use the SQA data booklet to help you with these questions.

1. The ion-electron equation shows what happens to the sulfite ions during the reaction.

    $SO_3^{2-}(aq) + H_2O(\ell) \rightarrow SO_4^{2-}(aq) + 2H^+(aq) + 2e^-$

    State what kind of reaction is taking place.

    ....

2. Write an ion-electron equation to show what happens to the iodine ($I_2(aq)$) during the reaction.

    ....

3. State the path and direction of electron flow.

    ....

Exam-style question

1. A student set up the following cell.

Figure 4.3

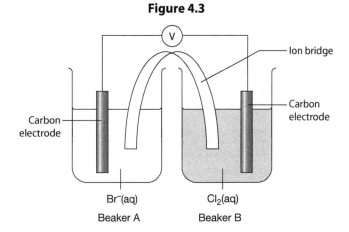

The equations for the reactions taking place at each electrode are shown below.

Beaker A: $2Br^-(aq) \rightarrow Br_2(aq) + 2e^-$

Beaker B: $Cl_2(aq) + 2e^- \rightarrow 2Cl^-(aq)$

a. Name the type of reaction happening in Beaker B. [1 mark]

....

b. Write the redox equation for the reaction. [1 mark]

....

c. On the diagram, draw an arrow to show the path and direction of electron flow. [1 mark]

Assignment support

Combining different metal half-cells with a copper half-cell can be used to investigate the effect of pairing different metals on the voltage obtained in an electrochemical cell. This would allow you to arrange the metals in an electrochemical series and use it to predict voltages which would be produced when different metals are combined. Other variables such as the surface area of the metal, temperature of the solutions and concentration of the electrolyte might also be investigated.

5 Making a soluble salt

Underlying chemistry

Insoluble bases can be reacted with acids to form soluble salts. The base can be an oxide, hydroxide or carbonate. A neutralisation reaction takes place. If excess of the base is added to the acid the unreacted base can be filtered off and the soluble salt solution is left behind. The water can then be evaporated from the salt solution leaving crystals of the salt behind.

Learning outcomes

- Carry out a neutralisation reaction between an acid and excess of an insoluble base.
- Carry out a filtration to separate the excess base from the salt solution.
- Evaporate the water from the salt solution by heating the solution with a Bunsen flame.
- Name the salt formed when an acid reacts with a base.
- Write a balanced equation for the reaction between an acid and a base.

Aim

To prepare copper(II) sulfate crystals by reacting excess copper(II) oxide with dilute sulfuric acid.

Apparatus list

Your teacher will ensure that all the apparatus you need for the experiment can be found in the laboratory. You can use this list to check that you have everything you need.

- 1.0 mol l^{-1} sulfuric acid
- copper(II) oxide
- spatula
- glass rod
- 100 cm^3 **beaker**
- Bunsen burner
- tripod with gauze

- heat-proof mat
- **filter funnel**
- **filter paper**
- 250 cm^3 **conical flask**
- **evaporating basin**
- crystallising dish

Safety notes

Your teacher will go over general safety rules with you and precautions which should be taken when carrying out this experiment.

In particular you should be aware of the following:

- Sulfuric acid is an irritant.
- Copper(II) oxide and copper(II) sulfate are harmful.

Experimental precautions

- Add the copper(II) oxide in small amounts to speed up the reaction.
- The acid can be gently **warmed** to speed up the reaction, not boiled.

Method

1. Pour sulfuric acid into the beaker, up to the 25 cm³ (ml) mark.

2. Use a spatula to add a small amount of the copper(II) oxide to the acid.

3. Use the glass rod to stir the mixture.

   (**Note**: if the copper(II) oxide is too slow to react ask your teacher if you should **warm** the mixture, using a Bunsen burner.)

4. Keep adding copper(II) oxide and stirring until the solution formed has a definite blue colour and there is excess copper(II) oxide in the bottom of the beaker.

Figure 5.1

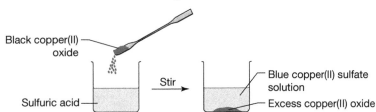

5. Make the filter paper into a cone shape and place it in a filter funnel.

6. Place the filter funnel in the conical flask.

7. Carefully pour the contents of the beaker into the filter funnel.

8. The copper(II) sulfate solution will collect in the conical flask.

9. The excess copper(II) oxide collects in the filter paper, which can now be discarded.

10. Pour the copper(II) sulfate solution into an evaporating basin.

11. Place the evaporating basin on a tripod stand with gauze and heat the basin gently.

Figure 5.2

12. Once the volume of the solution has reduced by about half, switch the Bunsen burner off and leave the evaporating dish to cool.

13. Once the evaporating dish has cooled, pour the contents into a crystallising dish and leave the dish in a warm place overnight.

Record your results

1. Draw the shape of two of the crystals which formed and colour them in.

2. Draw a labelled sectional (line) diagram showing the assembled apparatus you used to filter the mixture.

Check your understanding

1. **a.** Complete the word equation:

acid + base → .... + ...

b. Name the type of reaction taking place.

....

2. Write the chemical equation for sulfuric acid reacting with copper(II) oxide.

....

3. **a.** State what is meant by 'excess' copper(II) oxide.

....

b. State how you could tell when all the acid had reacted with the copper(II) oxide.

....

4. Copper carbonate could have been used to make the copper(II) sulfate instead of copper(II) oxide.

a. Write a chemical equation for the reaction.

....

b. State what you would see during the reaction which you would not see when using copper(II) oxide.

....

5. A soluble salt is also produced when an alkali is neutralised by an acid in a titration.

State one advantage using an insoluble base to make a soluble salt has over carrying out a titration.

....

....

Exam-style question

1. Magnesium is important for the normal functioning of cells, nerves, muscles and bones. Magnesium chloride is a mineral supplement used to prevent and treat low amounts of magnesium in the blood.

   a. Magnesium chloride can be made in the laboratory by reacting magnesium carbonate with hydrochloric acid.

      i. The equation for the reaction is:

      $MgCO_3(s) + HCl(aq) \rightarrow MgCl_2(aq) + CO_2(g) + H_2O(\ell)$

      Balance the equation. [1 mark]

      ii. Name the type of reaction taking place. [1 mark]

      ....

   b. Outline how you would carry out the experiment in the laboratory to produce a pure dry sample of magnesium chloride. [3 marks]

   ....

   ....

   ....

   ....

   ....

   ....

6 Determination of E_h

Underlying chemistry

When alcohols undergo a combustion reaction (burn) they release heat energy. This means alcohols can be used as fuels. The quantity of heat energy released (E_h) can be determined experimentally and calculated using the relationship $E_h = cm\Delta T$. The chemical products of combustion of alcohols are carbon dioxide and water. When ethanol burns in a plentiful supply of oxygen, carbon dioxide and water are the only products. No poisonous carbon monoxide is formed. Ethanol can be produced from renewable carbohydrate sources such as sugar cane, unlike petrol, which is produced from oil.

Learning outcomes

- Heat a known volume of water using the flame produced when ethanol is burned and measure the rise in temperature (ΔT) of the water.
- Use a balance to weigh accurately.
- Use the results from the experiment to calculate the heat energy released when 1·0 g of ethanol burns, using the relationship $E_h = cm\Delta T$.

Aim

Calculate the amount of heat energy produced when 1·0 g of ethanol burns.

Apparatus list

Your teacher will ensure that all the apparatus you need for the experiment can be found in the laboratory. You can use this list to check that you have everything you need.

- alcohol burner (containing ethanol) and lid
- heat-proof mat
- **electronic balance**
- clamp and stand
- copper **beaker** (or other metal)
- stirring **thermometer**
- draught shield
- 100 cm³ **measuring cylinder**

Safety notes

Your teacher will go over general safety rules with you and precautions which should be taken when carrying out this experiment.

In particular you should be aware of the following:

- Ethanol is very flammable.
- Take care when handling the alcohol burner as the ethanol flame is very hot.
- The copper beaker will get hot so do not touch it when it has been heated by the flame.

Experimental precautions

- There needs to be about a 3 cm gap between the bottom of the beaker and the wick of the alcohol burner.
- The temperature rise of the water need not be exactly 20°C but the reading must be accurate.
- Make sure the lid is on the alcohol burner when it is weighed both before and after it is used.

Method

1. Weigh the alcohol burner with the lid on and note the mass in your results table.

2. Place the alcohol burner on the heat-proof mat.

3. Clamp the copper beaker so that the bottom of the beaker is about 3 cm from the wick of the burner.

4. Measure out 100 cm³ of water using the measuring cylinder and pour it into the beaker.

5. Take the temperature of the water and note it in your results table.

6. Remove the lid from the burner and light the wick.

Figure 6.1

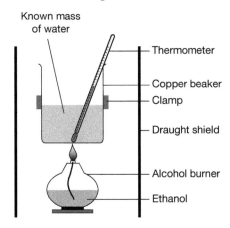

7. Stir the water with the thermometer.

8. When the temperature has risen by about 20°C move the burner from under the beaker and place the lid over the wick to extinguish the flame.

9. Continue to stir the water and note the highest temperature reached. Record the temperature in your results table.

10. Reweigh the alcohol burner and note the weight in your results table.

Record your results

1. Present your results in a table with appropriate headings.

2. Use your results to calculate E_h, in kJ.

Use $E_h = cm\Delta T$

where c = specific heat capacity of water (4.18 kJ kg^{-1} °C^{-1})

m = mass of water (kg)

ΔT = change in the temperature of the water (°C)

3. Calculate E_h, in kJ, for 1.0 g of alcohol.

Check your understanding

1. State why a copper beaker is used instead of a glass one.

....

2. State why the draught shield is used.

....

3. Suggest why the water should be stirred after the burner has been extinguished.

....

4. Explain why the experimental value obtained for the combustion of 1 mol of ethanol is much less than the data book value.

....

5. Suggest why it is important to keep the lid on the burner when it is not lit.

....

6. Data obtained from an experiment to find the specific heat capacity of a liquid is shown in the table.

Measurement	Result
Enthalpy change (kJ)	2·55
Mass of liquid heated (g)	50
Initial temperature of liquid (°C)	22·9
Final temperature of liquid (°C)	44·2

Calculate the specific heat capacity of the liquid, in kJ kg^{-1} °C^{-1}.

Exam-style question

1.   Alcohols burn, releasing heat energy.

    a. State the term used to describe all chemical reactions that release heat energy.   [1 mark]

    ....

    b. A student investigated the amount of energy released when an alcohol burns using the apparatus shown.

Figure 6.2

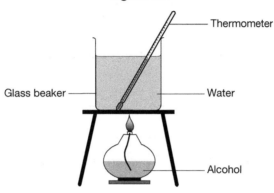

The student recorded the following data.

Mass of alcohol burned	1 g
Volume of water	200 cm³
Initial temperature of water	15°C
Final temperature of water	35°C

    i.   Calculate the energy released, in kJ.

         Show your working clearly.   [3 marks]

    ii. Suggest one improvement to the student's investigation.   [1 mark]

    ....

c. The table gives information about the amount of energy released when one mole of some alcohols is burned.

Name of alcohol	Energy released when one mole of alcohol is burned (kJ)
methanol	726
ethanol	1367
propan-1-ol	2021
butan-1-ol	2676

    i.   Write a statement linking the amount of energy released to the number of carbon atoms in the alcohol molecule. [1 mark]

....

....

   ii.   Predict the amount of heat released, in kJ, when one mole of pentan-1-ol is burned. [1 mark]

....

Assignment support

The experimental technique could be used to compare the amount of heat produced when 1.0 g of different alcohols are burned. A link could be made between the number of carbon atoms in the molecule and the amount of heat produced when they burn.

7 Electrolysis of copper(II) chloride solution

Underlying chemistry

Electrolysis is the decomposition (breaking down) of an ionic compound into its elements by passing electricity through a solution or melt (liquid). A direct current (d.c.) is used so that the products of electrolysis at the electrodes can be identified. Positive ions are attracted to the negative electrode where they gain electrons and form atoms. Negative ions are attracted to the positive electrode where they lose electrons and form atoms. Electrolysis is used in industry to extract reactive metals from their compounds.

Learning outcomes

- Carry out an electrolysis and identify the products at the electrodes.
- Write ion-electron equations for the reactions taking place at the electrodes.
- Identify the electrodes at which oxidation and reduction reactions take place.

Aim

Carry out the electrolysis of copper(II) chloride solution and identify the products at the electrodes.

Apparatus list

Your teacher will ensure that all the apparatus you need for the experiment can be found in the laboratory. You can use this list to check that you have everything you need.

- two **carbon electrodes**
- plastic D-piece for holding the electrodes
- 1·0 mol l^{-1} copper(II) chloride solution
- 100 cm^3 **beaker**

- d.c. supply
- two leads
- two crocodile clips

Safety notes

Your teacher will go over general safety rules with you and precautions which should be taken when carrying out this experiment.

In particular you should be aware of the following:

- Make sure you carry out the electrolysis in a well-ventilated area.
- Do not try to sniff the gas produced when trying to identify the gas, waft the gas produced towards you using your hand.

Experimental precautions

- The carbon rods are very brittle so take care when handling them.
- Make sure the carbon rods are not touching when the d.c. supply is on.
- Slowly increase the voltage until a steady stream of gas is produced.

Method

1. Pour the copper(II) chloride solution into the beaker, up to the 75 cm^3 (ml) line.
2. Place the D-piece on the beaker.
3. Put the carbon rods through the holes in the D-piece.

   (**Note**: If you do not have a D-piece, place the carbon rods in the solution but make sure they are not touching.)
4. Connect a crocodile clip to each of the leads.
5. Connect a crocodile clip and lead to each electrode.
6. Connect the other end of each lead to the d.c. supply.
7. Set the voltage on the d.c. supply to 2 V and observe whether anything is happening at the electrodes.
8. Increase the voltage to 4 V if there is no change observed at the electrodes.
9. Note what you observe at each electrode.
10. Try to identify the product at the positive electrode by its smell.
11. Switch off the voltage once you have made your observations.

Record your results

Draw a labelled sectional (line) diagram of the electrolysis apparatus, showing clearly the positive and negative electrodes and the products at each electrode.

Check your understanding

1. Explain why a d.c. supply is used.

....

....

....

2. **a. i.** Write an ion-electron equation for the reaction happening at the positive electrode.

....

    ii. Name the type of reaction taking place at the positive electrode.

....

  b. i. Write an ion-electron equation for the reaction happening at the negative electrode.

....

    ii. Name the type of reaction taking place at the negative electrode.

....

  c. Write the redox equation.

....

3. **a.** State why the compound being electrolysed has to be in a solution or in the liquid state.

....

  b. Explain why covalent compounds cannot be decomposed by electrolysis.

....

Exam-style question

1. Aluminium can be extracted from naturally occurring metal compounds such as bauxite. Bauxite is refined to produce aluminium oxide. Electrolysis of molten aluminium oxide produces aluminium and oxygen gas.

  a. Explain why a d.c. supply **must** be used. [1 mark]

....

  b. State why the aluminium oxide **must** be molten. [1 mark]

....

  c. Write the ion-electron equation for the reaction taking place at the negative electrode.

[1 mark]

....

  d. During electrolysis, oxide ions lose electrons at the positive electrode to form oxygen gas.

    Name the type of chemical reaction taking place. [1 mark]

....

Answers

1 Monitoring the rate of reaction

Check your understanding

1. All variables other than the one which is being changed must remain the same to make the comparison fair.
2. **a.** The rate of reaction decreases.
   b. The reactants are getting used up. OR The concentration of the acid is decreasing. OR The surface area of the marble chips is decreasing.
3. When one of the reactants is used up the reaction will stop.
4. $CaCO_3 + 2HCl \longrightarrow CaCl_2 + H_2O + CO_2$
5. Weigh the marble chips in both experiments to ensure the same mass is used.

Exam-style question

1. **a.** There is more than enough magnesium to react with all of the acid. [1]
   b. Hold a lighted splint at the mouth of the test tube [1] and the gas burns with a 'pop'. [1]

   c. i. Average rate $= \dfrac{\text{change in volume}}{\text{change in time}}$ [1]

   $$= \frac{70 - 46}{30 - 10}$$

   $$= \frac{24}{20}$$

   $$= 1.2 \text{ cm}^3 \text{ s}^{-1}$$

   [1 for answer, 1 for units]

   (The relationship $\text{rate} = \dfrac{\Delta \text{quantity}}{\Delta t}$ can be found in the SQA data booklet.)

   ii. The number of moles of hydrogen ions in sulfuric acid (H_2SO_4) is twice that of hydrochloric acid (HCl). [1]
   d. The reaction would be faster. [1]

2(a) Making a standard solution

Check your understanding

1. The mass given is only a guide to the amount needed. The actual mass recorded has to be accurate so the concentration of the solution can be worked out accurately.
2. The tare function means that the filter paper can be placed on the balance and the balance then reset to zero so that the mass of the substance being weighed can be read directly from the balance.
3. To ensure that all of the sodium hydroxide solution is transferred to the volumetric flask.
4. Tap water might contain dissolved substances which could have an effect on the concentration of the sodium hydroxide solution.

Exam-style question

1. **a.** Solution of accurately known concentration. [1]
   b. Step 1: Calculate the number of moles using the relationship $n = m$/GFM.
   $n = 5.3/106 = 0.05$ (mol) [1]

(**Note:** The relationship $n = m$/GFM can be found in the SQA data booklet.)
Step 2: Calculate the concentration using the relationship $C = n/V$. $C = 0.05/0.5 = 0.1$ mol l^{-1}
[1 for calculation, 1 for units]
(**Note:** The relationship $n = CV$ can be found in the SQA data booklet. From this $C = n/V$ can be obtained.)

2(b) Using a standard solution in a titration

Results

1. Pink to colourless
2. Your table should be similar to the one below, with your own results.

Titration	Initial reading (cm³)	Final reading (cm³)	Total volume added (cm³)
1			
2			
3			
4 etc.			

3. The first titre is often not very accurate so is not used to work out the average.
   The two titres within 0.2 cm³ of each other should be added and divided by two.

4. Use $C_1 = \dfrac{C_2 \times V_2 \times n_1}{n_2 \times V_1}$.

   Unit of concentration is mol l^{-1}.

Check your understanding

1. The total volume added is obtained by subtracting the initial volume from the final volume.
2. To ensure the solutions get mixed so they react.
3. So the colour change at the endpoint can be seen more clearly.
4. The acid has completely neutralised the alkali.
5. With universal indicator there would be a number of colour changes as the acid is added rather than a single sharp colour change at the endpoint.
6. The first titre is often not accurate enough to be used when working out the average.

Exam-style questions

1. **a.** They are concordant – they are within 0.2 cm³ of each other. [1]

   b. $C_1 = \dfrac{C_2 \times V_2 \times n_1}{n_2 \times V_1}$

   $$= \frac{0.05 \times 20.0 \times 1}{13.2 \times 1}$$ [1]

   $C_1 = 0.076$ mol l^{-1} [1 for answer, 1 for units]

2.

a. An indicator [1]

b. Burette [1]

c. $C_1 = \dfrac{C_2 \times V_2 \times n_1}{V_1 \times n_2}$

$= \dfrac{0{\cdot}1 \times 25{\cdot}0 \times 2}{24{\cdot}8 \times 1}$ [1]

$C_1 = 0{\cdot}2 \text{ mol l}^{-1}$ [1 for answer, 1 for units]

3(a) Identifying ions – flame tests

Check your understanding

1. A blue flame is the hottest and is almost invisible.
2. So that there is no contamination from a previous sample.
3. Tap water could contain ions which could show up in the flame and so give a false result.

Exam-style question

1. **a.** Barium [1]

   b. Carry out a flame test on a known sample containing lithium to get the colour of the flame then do the test on the unknown compound and compare the colours. [1]
   Repeat the process with the known sample containing strontium. [1]

3(b) Identifying ions – forming a precipitate

Check your understanding

1. The precipitate is the solid formed when two solution are mixed.
2. So that other ions are removed so they will not react with the silver nitrate.
3. The silver fluoride formed is soluble in water.
4. **a.** $NaCl(aq) + AgNO_3(aq) \longrightarrow AgCl(s) + NaNO_3(aq)$

   b. Silver bromide

3(c) Identifying the ions in an unknown metal halide

Exam-style questions

1. **a.** **i.** Silver iodide [1]

   ii. Lithium carbonate OR lithium oxide [1]

   b. **i.** The precipitates could be a different colour. [1]

   ii. Carry out a flame test on both compounds [1]
   and they should produce the same colour of flame (red). [1]

2. **a.** The test could be used to identify aluminium ions [1]

   but could not tell if calcium or magnesium ions were present as both give the same result. [1]

   b. **i.** $CuSO_4(aq) + 2NaOH(aq) \longrightarrow Cu(OH)_2(s) + Na_2SO_4(aq)$ [1]

   ii. Copper hydroxide OR copper(II) hydroxide [1]

   c. $Na^+(aq) + Cl^-(aq)$ [1]

   d. Carry out flame tests. [1]
   Each would produce a different colour of flame. [1]

4(a) Electrochemical cells – cell with metal electrodes

Check your understanding

1. The circuit was not complete so there was no electricity flowing.
2. Electrolytes
3. **a.** Ion or salt bridge

   b. Allows ions to move between the solutions.

4. **a.** $Zn(s) \longrightarrow Zn^{2+}(aq) + 2e^-$

   b. $Cu^{2+}(aq) + 2e^- \longrightarrow Cu(s)$

   c. $Zn(s) + Cu^{2+}(aq) \longrightarrow Zn^{2+}(aq) + Cu(s)$

   d. Electrons flow from the zinc to the copper through the wire.

Exam-style question

1. **a.** Electrons flow from the magnesium to the silver, through the wire. [1]

   b. So that ions can move between the beakers. [1]

   c. $Ag^+(aq) + e^- \longrightarrow Ag(s)$ [1]

   d. The reading would decrease since the gap between zinc and silver in the electrochemical series is smaller than between magnesium and silver. [1]

4(b) Electrochemical cells – cell with non-metal electrodes

Check your understanding

1. Oxidation
2. $I_2(aq) + 2e^- \longrightarrow 2I^-(aq)$
3. From the sulfite solution to the iodine, through the wire.

Exam-style question

1. **a.** Reduction [1]

   b. $2Br^-(aq) + Cl_2(aq) \longrightarrow Br_2(aq) + 2Cl^-(aq)$ [1]

   c. From Beaker A to Beaker B, through the wire. [1]

Practical 5 Making a soluble salt

Check your understanding

1. **a.** acid + base $\longrightarrow$ water + a salt

   b. Neutralisation

2. $H_2SO_4 + CuO \longrightarrow CuSO_4 + H_2O$

3. **a.** More than enough to react with all of the acid.

   b. There would be unreacted copper(II) oxide lying in the bottom of the beaker.

4. **a.** $H_2SO_4 + CuCO_3 \longrightarrow CuSO_4 + CO_2 + H_2O$

   b. Bubbles of carbon dioxide gas.

5. Using a titration means that an indicator has to be used so that the endpoint of the reaction can be detected. The titration then has to be repeated without the indicator.

Exam-style question

1. **a.** **i.** $MgCO_3(s) + 2HCl(aq) \longrightarrow MgCl_2(aq) + CO_2(g) + H_2O(\ell)$ [1]

   ii. Neutralisation [1]

   b. Add the magnesium carbonate to the hydrochloric acid until no more carbonate reacts. [1]
   Filter the mixture and collect the magnesium chloride solution. [1]

Heat the solution to evaporate off some of the water and leave the rest in a crystallising dish to form crystals. [1]

6 Determination of E_h

Check your understanding

1. Metal conducts the heat to the water better than glass does.
2. To reduce heat lost to the surrounding air.
3. So that there is time for all the heat in the copper beaker to be transferred to the water.
4. So much heat is lost to the surroundings and not transferred to the water.
5. So that the ethanol does not evaporate leading to inaccuracies in the mass of ethanol burned.
6. $E_h = cm\Delta T$ so $c = E_h/m\Delta T$
   $= 2{\cdot}55/(0{\cdot}05 \times 21{\cdot}3)$
   $= 2{\cdot}39$ (kJ kg^{-1} °C^{-1})

Exam-style question

1. **a.** Exothermic [1]
   b. i $E_h = cm\Delta T$ [1]
   $= 4{\cdot}18 \times 0{\cdot}2 \times 20$ [1]
   $= 16{\cdot}72$ (kJ) [1]
   (Remember: The relationship $E_h = cm\Delta T$ and the value of c (the specific heat capacity of water) can be found in the SQA data booklet.)
   ii. Use a metal beaker instead of a glass beaker OR Use a draught shield OR Stir the water before taking a final reading. [1]
   c. i. As the number of carbon atoms increases the amount of energy released increases. [1]
   ii. Between −3276 and −3376 (actual = −3329) [1]

7 Electrolysis of copper(II) chloride solution

Results

Figure A.1

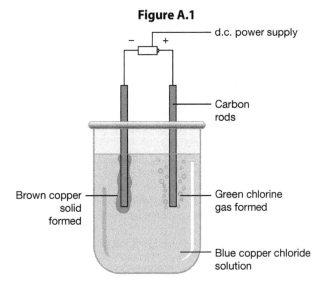

Check your understanding

1. With a d.c. supply one electrode is positive and the other negative. Negative ions are attracted to the positive electrode and positive ions are attracted to the negative electrode. The products formed at each electrode can be identified.
2. **a.** i. $2Cl^-(aq) \longrightarrow Cl_2(g) + 2e^-$
   ii. Oxidation
   b. i. $Cu^{2+}(aq) + 2e^- \longrightarrow Cu(s)$
   ii. Reduction
   c. $2Cl^-(aq) + Cu^{2+}(aq) \longrightarrow Cl_2(g) + Cu(s)$
3. **a.** So that the ions are free to move to the electrodes.
   b. There are no ions in a covalent compound.

Exam-style question

1. **a.** So that the products can be identified. [1]
   b. So that the ions can move to the electrodes. [1]
   c. $Al^{3+} + 3e^- \longrightarrow Al$ [1]
   d. Oxidation [1]

Notes

Notes